Introduction

Introduction

Arab inventors have played a significant role in shaping the world we live in today. During the Islamic Golden Age, which lasted from the 8th to the 14th centuries, the Arab world was a hub of scientific and intellectual activity. Many great thinkers and inventors emerged during this time, producing groundbreaking works in fields such as mathematics, astronomy, chemistry, medicine, and engineering. These inventors were not only innovators but also scholars who contributed to the advancement of human knowledge, and their discoveries have had a lasting impact on the world.

Arab inventors were known for their creativity, ingenuity, and problem-solving skills. They developed new technologies, invented machines, and discovered scientific principles that changed the course of history. Their inventions included things like the astrolabe, which was used for navigation and astronomy, the windmill, which was used for grinding grain and pumping water, and the camera obscura, which was used for observing and documenting the world.Many Arab inventors were also polymaths who excelled in multiple fields of study. They combined their knowledge of mathematics, science, and engineering to create inventions that were both practical and innovative. Some of the most famous Arab inventors include Al-Khwarizmi, the father of algebra, Ibn al-Haytham, the father of optics, and Jabir ibn Hayyan, the father of chemistry.

Today, the legacy of Arab inventors continues to inspire and motivate new generations of innovators. The world owes a great debt to these inventors, whose works have made life easier, more comfortable, and more interesting for people all over the globe.

Chapter 1: Al-Khwarizmi - The Father of Algebra

Al-Khwarizmi was born in the city of Khiva, which is now part of Uzbekistan, around 780 CE. He was a Persian mathematician, astronomer, and geographer who lived in the Arab world during the Islamic Golden Age. Al-Khwarizmi is known as the father of algebra, and his book "Al-Kitab al-mukhtasar fi hisab al-jabr wal-muqabala" (The Compendious Book on Calculation by Completion and Balancing) introduced the concept of algebraic equations and systematic solutions for them.

Before Al-Khwarizmi, mathematical problems were mainly solved geometrically, and equations were described in words rather than using symbols. Al-Khwarizmi's book introduced the concept of using letters to represent unknown quantities and provided a systematic method for solving linear and quadratic equations.

This concept revolutionized mathematics and had a significant impact on the development of science and engineering. Al-Khwarizmi's work on algebra was not only significant in its own right but also paved the way for further advancements in mathematics. His work on equations was essential for solving problems in fields such as physics, engineering, and economics, where equations are used extensively to model and predict real-world phenomena. Al-Khwarizmi's work on algebra also influenced the development of other areas of mathematics, such as trigonometry and calculus.

In addition to his work on algebra, Al-Khwarizmi made significant contributions to trigonometry and introduced the concept of the algorithm.

His work on trigonometry included the development of tables of sines and tangents, which were used for astronomical calculations. Al-Khwarizmi's work on algorithms included the development of a systematic method for solving linear equations that is still used today.

The term "algorithm" is derived from Al-Khwarizmi's name, and his work on algorithms had a significant impact on the development of computer science. His method for solving linear equations is the basis for the Gaussian elimination algorithm, which is used in a wide range of applications, including computer graphics and machine learning.

Al-Khwarizmi's legacy extends beyond his work in mathematics and computer science. He was also an accomplished geographer and astronomer, and his work on geography and cartography was instrumental in expanding the knowledge of the world's geography. He developed a method for calculating the direction of Mecca from any location, which is still used by Muslims today to determine the direction of prayer.

Overall, Al-Khwarizmi's contributions to mathematics, science, and engineering were fundamental to the development of these fields. His work laid the foundation for modern algebra and algorithms and influenced the development of computer science and other areas of mathematics. Al-Khwarizmi's legacy continues to inspire and inform new generations of scholars and researchers.

Al-Khwarizmi is known for developing algebra and introducing systematic solutions for algebraic equations in his book "Al-Kitab al-mukhtasar fi hisab al-jabr wal-muqabala" (The Compendious Book on Calculation by Completion and Balancing). Here are a few examples of equations that Al-Khwarizmi worked with in his book:

1. Solve the equation $2x + 5 = 11$:
   - Al-Khwarizmi's method involves "completing the square," which means adding a constant to both sides of the equation to make it factorable. In this case, we can add 2.5 to both sides: $2x + 5 + 2.5 = 11 + 2.5$ $2x + 7.5 = 13.5$
   - Now we can factor the left side of the equation as follows:
   - $(x + 1.875)^2 = 15.5625$

- Finally, we can take the square root of both sides to get the solution:
- $x + 1.875 = \pm 3.95$ $x = -1.875 \pm 3.95$ $x = 2.075$ or $x = -5.825$

1. Solve the equation $x^2 + 6x = 7$:
   - Al-Khwarizmi's method involves "completing the square" again. In this case, we can add 9 to both sides of the equation:
   - $$x^2 + 6x + 9 = 16$$
   - Now we can factor the left side of the equation as follows:
   - $$(x + 3)^2 = 16$$
   - Finally, we can take the square root of both sides to get the solution:
   -
   - $x + 3 = \pm 4$ $x = 1$ or $x = -7$

Chapter 2: Ibn al-Haytham - The Father of Optics

Ibn al-Haytham, also known as Alhazen, was a brilliant scientist and philosopher born in Basra, Iraq, in 965 CE. He is considered the father of optics, and his work on optics influenced the development of astronomy, mathematics, physics, and even the visual arts.

Ibn al-Haytham's most famous work is "Kitab al-Manazir" or "The Book of Optics," which is a comprehensive study of light and vision. The book was written in the 11th century and includes the first scientific explanation of the way in which the eye perceives images. It also describes the principles of reflection, refraction, and the camera obscura.

The camera obscura is a device that uses a pinhole to project an inverted image of the outside world onto a surface inside a dark room, and it was used by artists to create realistic images.

Ibn al-Haytham's work on optics had a profound impact on the development of the scientific method. He introduced the idea of using experiments to test scientific theories, and he emphasized the importance of observation and empirical evidence. His work on optics was also important in the development of modern optics and helped lay the foundation for the study of light and color.

In addition to his work on optics, Ibn al-Haytham made significant contributions to other fields of science and mathematics. He worked on the development of trigonometry, geometry, and algebra. He also wrote several treatises on astronomy, including a critique of the Ptolemaic model of the universe, which had been widely accepted at the time.

Ibn al-Haytham's impact on the visual arts cannot be understated either. His work on optics and the camera obscura had a significant impact on artists, particularly during the Renaissance period. Artists such as Leonardo da Vinci and Albrecht Durer used the camera obscura to create realistic images and to study the principles of light and color.

Ibn al-Haytham's legacy continues to inspire and inform new generations of scientists, mathematicians, and artists. His work on optics laid the foundation for modern optics and had a significant impact on the development of the scientific method.

His contributions to mathematics and astronomy were also significant and helped lay the foundation for these fields.

.Overall, Ibn al-Haytham's contributions to science and the visual arts were fundamental to the development of these fields, and his legacy continues to be celebrated and studied today.

Chapter 3: Jabir Ibn Hayyan - The Father of Chemistry

Jabir Ibn Hayyan, also known as Geber in the Western world, was an alchemist, chemist, and philosopher born in Tus, Iran, in the early 8th century. He is considered the father of chemistry and is credited with laying the foundations for modern chemistry. Jabir Ibn Hayyan was a polymath, and his work encompassed many fields, including alchemy, chemistry, physics, philosophy, and medicine.

Jabir Ibn Hayyan's most significant contribution to chemistry was the development of the scientific method in the field of alchemy. Before Jabir Ibn Hayyan, alchemy was more of a mystical practice, focused on the transmutation of metals and the search for the philosopher's stone.

However, Jabir Ibn Hayyan introduced the scientific method to alchemy, emphasizing observation, experimentation, and the importance of documentation.

Jabir Ibn Hayyan also made significant contributions to the study of acids, bases, and salts. He was the first to describe many chemical processes and substances, including nitric acid, hydrochloric acid, and aqua regia. He also developed the concept of the reduction of metals and the preparation of pure metals, which became essential in the development of modern metallurgy.

In addition to his work on chemistry, Jabir Ibn Hayyan also made significant contributions to the fields of medicine and pharmacology.

He wrote several treatises on medicine, including a book on the preparation of drugs and remedies. His work in pharmacology focused on the use of plants and minerals in the preparation of medicines, and his work on distillation and sublimation led to the development of the modern pharmaceutical industry.

Jabir Ibn Hayyan's impact on the development of modern chemistry is undeniable. His work laid the foundation for the scientific method in chemistry and had a significant impact on the development of the modern chemical industry. His work on acids, bases, and salts was instrumental in the development of modern chemistry, and his work in metallurgy helped advance the field of materials science.

Jabir Ibn Hayyan's legacy continues to inspire and inform new generations of scientists, chemists, and alchemists. His contributions to the development of the scientific method in alchemy and chemistry were fundamental to the development of these fields, and his work in medicine and pharmacology helped lay the foundation for modern medicine. Overall, Jabir Ibn Hayyan's contributions to science were significant and far-reaching, and his legacy continues to be celebrated and studied today.

Chapter 4: Ibn Sina - The Prince of Physicians

Ibn Sina, also known as Avicenna, was a Persian polymath who lived in the Arab world during the Islamic Golden Age. He is known as the prince of physicians and is considered one of the most significant figures in the history of medicine.

Ibn Sina was born in 980 CE in Bukhara, which is now in Uzbekistan. He showed an early aptitude for learning, and by the age of 16, he had already mastered several fields of study, including mathematics, logic, and metaphysics. He was also interested in medicine and began to study the works of earlier Greek and Arabic physicians.

Ibn Sina's interest in medicine led him to pursue a career in the field. He served as a physician to several rulers, including the Samanid ruler Nuh ibn Mansur and the Buyid ruler Shams al-Dawla. He also wrote many works on medicine, including his most famous work, the "Canon of Medicine."

The "Canon of Medicine" is a medical encyclopedia that became the standard textbook in European universities for centuries. It is divided into five books, which cover a wide range of medical topics, including anatomy, physiology, pathology, and pharmacology.

The book was written in Arabic, but it was translated into Latin in the 12th century and became an essential part of the medical curriculum in Europe.

One of the significant contributions of the "Canon of Medicine" was its emphasis on the importance of observation and diagnosis. Ibn Sina believed that a physician must first understand the symptoms of a disease and then use that understanding to make a diagnosis. He also stressed the importance of preventive medicine, including proper diet and exercise, as a way to maintain good health.

Ibn Sina made several other significant contributions to the field of medicine. He was the first to describe meningitis, and he also wrote extensively on the treatment of infectious diseases. He introduced several new drugs, including potassium nitrate, which was used to treat heart disease, and introduced the concept of quarantine to prevent the spread of infectious diseases.

Ibn Sina is most well-known for his contributions to the field of medicine, so he did not develop mathematical equations in the same way as Al-Khwarizmi did.

However, he did make significant contributions to the development of physics and metaphysics, and some of his ideas can be expressed mathematically.

One of Ibn Sina's most important ideas was his distinction between essence (mahiyya) and existence (wujud) in metaphysics. He argued that every object has both an essence and an existence, and that the existence of an object is contingent upon a cause. This can be expressed mathematically as:
Existence = Cause + Essence

This equation suggests that an object's existence is dependent on both its essential properties and an external cause or condition.

Another important idea in Ibn Sina's philosophy was his understanding of motion and change. He believed that motion was a continuous process that could be broken down into smaller, indivisible units. He also argued that every object had an inherent tendency towards either rest or motion, which could be expressed mathematically as:

$$F = ma$$

This equation, known as Newton's second law of motion, suggests that the force (F) acting on an object is equal to its mass (m) times its acceleration (a). Ibn Sina did not develop this equation himself, but his ideas about motion and change helped to lay the foundation for its development by later scientists.

In addition to his work in medicine, Ibn Sina also made significant contributions to philosophy, mathematics, and astronomy. He wrote several works on philosophy, including the "Book of Healing," which discussed logic, metaphysics, and ethics. He also wrote on mathematics and introduced the concept of the "fourth dimension" in geometry. Finally, he made significant contributions to astronomy, including his work on the movement of the planets.

In conclusion, Ibn Sina was a remarkable figure whose contributions to medicine and other fields continue to be recognized today. His emphasis on observation and diagnosis, his introduction of new drugs, and his work on preventive medicine were significant contributions that continue to influence the field of medicine.

His works in philosophy, mathematics, and astronomy also make him one of the most significant figures of the Islamic Golden Age.

Chapter 5: Al-Jazari - The Master of Automata

Al-Jazari was a renowned Turkish inventor, mechanical engineer, artisan, and polymath who lived in the Arab world during the 13th century.

He is famously known as the master of automata and is famous for his elaborate and sophisticated mechanical devices, including clocks, fountains, musical automata, and robots.

Al-Jazari was born in the city of Cizre, in present-day Turkey, in the year 1136. His full name was Badi' al-Zaman Abu al-'Izz Isma'il Ibn al-Razzaz al-Jazari, but he is commonly known as Al-Jazari. His father was an artisan who worked in a waterworks, which may have influenced Al-Jazari's interest in machines and automata.

Al-Jazari served as an engineer in the court of the Artuqid dynasty, who ruled the region at the time. He was commissioned by the Artuqids to design and construct sophisticated mechanical devices for their palaces and public spaces.

Over his lifetime, Al-Jazari wrote several books on mechanics and automata, including the famous "The Book of Knowledge of Ingenious Mechanical Devices" (Kitab fi ma'rifat al-hiyal al-handasiyya).

This book describes a variety of mechanical devices, including clocks, fountains, musical automata, and robots. The book also explains the principles of mechanics, including gears, pulleys, and levers.

One of the most famous inventions described in the book is the Elephant Clock, which was a complex water-powered clock that included several automata, including an elephant and a bird.

The Elephant Clock consisted of a large wooden elephant that stood on a platform. The elephant had a howdah, or carriage, on its back, which housed a large clock.

On top of the clock, there was a bird that flapped its wings and made sounds at certain times of the day. The elephant moved its head, trunk, and tail, and opened its mouth to reveal a set of mechanical musicians who played music.

Al-Jazari's designs were not only sophisticated but also aesthetically pleasing

He used a variety of materials to make his machines, including brass, copper, iron, and silver. He also decorated his machines with calligraphy and intricate patterns, which made them not only functional but also works of art.

Al-Jazari's contributions to the field of automata and mechanics were significant and had a lasting impact on the world. His designs were studied and replicated by other engineers and artisans, and they inspired future generations of inventors, including Leonardo da Vinci.

Today, Al-Jazari is recognized as one of the most important figures in the history of technology and engineering.

Chapter 6: Abbas ibn Firnas - The First Flying Man

Abbas ibn Firnas was an Andalusian inventor, polymath, musician, and poet who lived in the Arab world during the 9th century. He was born in the city of Ronda in Spain, which was under Muslim rule at the time. Abbas was fascinated by science and technology from an early age and developed a passion for exploring the unknown.

Abbas ibn Firnas made significant contributions in various fields, including astronomy, mathematics, music, and poetry. However, he is best known for his attempts to fly, which have earned him a place in history as the first flying man.

Inspired by the ancient Greek legend of Icarus, Abbas began designing and building his own flying machine.

He spent many years studying birds, their movements, and the principles of aerodynamics. Abbas realized that the key to successful flight was not just flapping wings but also gliding.

He designed and built a pair of wings made of silk and eagle feathers, which he believed would enable him to soar through the air.

Abbas had a vision of soaring over the city and showing the people the wonders of the world from above.

Abbas ibn Firnas chose the tallest mountain in the region, the Sierra Nevada, for his first attempt at flying. He climbed to the top of the mountain and strapped on his wings.

Abbas flapped his wings vigorously and attempted to take off, but he was not successful. He fell off the mountain and injured himself.

Undeterred by his failure, Abbas continued to improve his flying machine. He made changes to the design of his wings and tried again. This time, he chose a tower in Cordoba for his attempt.

Abbas climbed to the top of the tower and strapped on his wings. He flapped his wings and took off, gliding for a short distance before landing safely.

Although Abbas ibn Firnas did not achieve sustained flight, he was the first person in history to build and test a flying machine.

His experiments and designs were groundbreaking and inspired future generations of inventors and scientists to continue exploring the possibilities of flight. Abbas ibn Firnas's contributions were not limited to aviation.

He was a multi-talented polymath who excelled in many fields. Abbas was an accomplished musician who played the lute and sang. He was also a poet who wrote beautiful and inspiring verses.

chapter 7: Idriss yazami

Idriss Yazami is a well-known Moroccan scientist, entrepreneur, and inventor who has made significant contributions to the fields of renewable energy and sustainable development in Africa. Here's a story about his life and work:

Idriss Yazami was born in Morocco and grew up in a modest family. From a young age, he was interested in science and technology, and he was fascinated by the possibilities of renewable energy.

After completing his education in Morocco, he went to France to pursue his studies in physics and chemistry, where he earned a Ph.D. in electrochemistry.

After completing his studies, Idriss Yazami returned to Morocco, where he began his career as a researcher and professor at Mohammed V University in Rabat.

In the 1980s, he became interested in the field of energy storage and began to focus his research on developing new and more efficient battery technologies.

In the 1990s, Idriss Yazami invented a new type of lithium-ion battery that could store more energy than previous models.

This invention was a major breakthrough, as it allowed for the development of smaller, more powerful batteries that could be used in a wide range of electronic devices, from cell phones to electric vehicles.

Yazami's invention revolutionized the field of energy storage and made it possible for renewable energy sources such as solar and wind power to be more widely adopted.

He continued to refine his battery technology over the years, and his work has been recognized with numerous awards and honors, including the prestigious Draper Prize in engineering.

In addition to his scientific work, Idriss Yazami has also been a successful entrepreneur, founding several companies that focus on developing sustainable energy solutions in Africa. His companies have helped to bring clean energy to remote communities and have created jobs and economic opportunities in the region.

Despite his many accomplishments, Idriss Yazami remains humble and dedicated to his work. He continues to conduct research and mentor young scientists, hoping to inspire the next generation to tackle the world's most pressing problems, including climate change and sustainable development.

www.ingramcontent.com/pod-product-compliance
Lightning Source LLC
Chambersburg PA
CBHW071116220526
45467CB00004B/1921